FLORA OF TROPICAL

CALLITRICHACEAE

INGA HEDBERG & OLOV HEDBERG

Creeping or submerged, monoecious, aquatic herbs with filiform stems and opposite, entire exstipulate leaves. Stems with axillary glandular scales; both stem and leaves with peltate hairs. Flowers axillary, either solitary or with one male and one female flower in the same axil; perianth absent; bracteoles 2, crescent-shaped or absent; male flower consisting of one stamen with a slender filament and a reniform anther; female flower consisting of a single carpel with a 4-celled ovary containing a single anatropous ovule in each locule; styles 2, filiform. Fruit schizocarpic, separating into 4 mericarps. Seeds with a fleshy endosperm.

A family with a single genus.

CALLITRICHE

L., Gen. Pl., ed. 5: 5 (1754); Schotsman in B.J.B.B. 55: 291–296 (1985) & in Bull. Mus. Hist. Nat. Paris, ser 4, 10: 3–17 (1988); O. Hedberg & I. Hedberg, Tropical African Callitriche, Biol. Skr. 54: 19–30 (2001)

Description as for the family.

An almost cosmopolitan genus with about 30 species, of which 4 are found in the Flora area.

The leaf shape and arrangement vary considerably with the habitat of the plant. Submerged plants often have lanceolate or linear leaves and long internodes, while in plants where the shoots reach the surface the internodes tend to be shorter and have rosettes of elliptic or spathulate leaves. Terrestrial plants tend to have ovate or suborbicular leaves. The adaptation of the genus to aquatic life has resulted in the reduction of several morphological features, so that only a few reliable characters remain which can be used for identification. The most useful character for identification is that of ripe fruits. Other taxonomically useful characters, albeit unsuitable for use in species' keys, are the morphology of pollen and the number of chromosomes. In pollen, the exine surface shows the *Croton* pattern, which is a reticulate pattern with processes (clavae) at the intersections of the walls (muri) (Erdtman, G., 1952, Pollen morphology and plant taxonomy: Angiosperms, p. 173). This pattern is present in all tropical African species. Chromosome counts from European material shows the somatic numbers between 6 and 40. The four tropical African species fall well within the range of variation. All tropical African material was earlier classified as *C. stagnalis* Scop., but recent taxonomic work has shown that several species can be recognized based on morphological, pollen and chromosomal differences.

1. Fruit distinctly winged · 2
   Fruit not winged · 3
2. Wings symmetrical around the mericarp; petiole mostly
   3-veined · 1. *C. oreophila*
   Wings asymmetric; petiole 1-veined at base · · · · · · · · · · · · 2. *C. anisoptera*

1

3. Larger leaves 10–16 mm long, 5(–8)-veined; petiole 1(–3)
-veined · 3. *C. keniensis*
Larger leaves usually 6–8 mm long, usually 3-veined; petiole
1-veined · 4. *C. vulcanicola*

1. **C. oreophila** *Schotsman* in B.J.B.B. 55: 222, t. 7 (1985); Hedberg & Hedberg in Fl. Ethiopia 2, 1: 427, fig. 52/1–8 (2000). Type: Rwanda, Gitovu, *Bequet* 780 (BR!, holo.)

Stem 10–35 cm long in water or 3–20 cm long on mud. Leaves submerged, floating or aerial, elliptic to spathulate, up to 11 mm long and 5 mm wide but often much smaller, 3–7-veined; petiole usually 3-veined. Flowers solitary, in the axils of leaves; bracteoles 2, falcate; stamens at dehiscence 1–2 mm, after dehiscence up to 7 mm long; anthers 0.5–0.6 mm long; styles up to 2 mm long, erect or spreading. Fruit suborbicular, 1.6–1.8 mm long and 1.6–2 mm wide, with a distinct, symmetrically winged mericarp. Pollen reticulum not as wide as the clavae; clavae irregularly triangular, surface uneven. Fig. 1/1–4 (p. 3).

UGANDA. Toro District: E slope of Ruwenzori, 14 Apr. 1948 *Hedberg* 802!; Kigezi District: Bukinda, Feb. 1948, *Purseglove* 2580! & ?Narosanje, Apr. 1945, *Purseglove* 1653!
KENYA. Aberdare, Kinangop, *Chandler* 2322!; Nanyuki District: West Mt Kenya Forest Station, 26 Dec. 1921, *Fries & Fries* 319!; Narok District: Enesambulai Valley, 15 Aug. 1970, *Greenway & Kanuri* 14557!
TANZANIA. Moshi District: Kilimanjaro, above Marangu, Feb. 1894, *Volkens* 1834!; Lushoto District: W Usambaras, 21 June 1953, *Drummond & Hemsley* 2968!; Mbeya District: Mbeya Mt, 13 Dec. 1962, *Richards* 17044!
DISTR. **U** 2; **K** 1, 3, 4, 6; **T** 2, 3, 7; W Tropical Africa, Cameroon, Congo (Kinshasa), Rwanda, Burundi, Sudan, Ethiopia
HAB. In stagnant or running water in ponds or small streams and on mud or moist soil; 1150–3250 m

SYN. [*C. stagnalis* sensu auctt. mult. e.g. De Wild., Pl. Bequaert. 2: 58 (1923) p.p.; Sam. in Notizbl. Bot. Gart. Berlin 85: 322 (1925) p.p.; W. Robyns, Fl. Spermat. Parc Nat. Albert 1: 484 (1948) p.p.; A.V.P.: 127 (1957) p.p.; A. Robyns, F.C.B. 7: 351 (1958) p.p.; U.K.W.F. ed. 2: 82 (1994) p.p., *non* Scop.]

2. **C. anisoptera** *Schotsman* in Bull. Mus. Hist. Nat. Paris, sér. 4, 10: 7 (1988). Type: Uganda, Toro District, Ruwenzori, Bujuku Valley near Bigo camp, 3400 m, *Hedberg* 398 (BR!, holo., BM, EA, K!, S!, UPS!, iso.)

Stem 5–25 cm long, prostrate, mat-forming. Leaves spathulate to elliptic, 4–6 mm long and 2–2.5 mm wide, 3–5-veined; petiole 1-veined. Flowers solitary, the male with 2 bracteoles; stamens 2–6 mm long, anthers ± 0.5 mm long; style 1.5–3.5 mm long. Fruit ± 1.3–1.5 mm long and 1.6–1.8 mm wide, variously and unequally winged with best developed wings in the distal part. Pollen reticulum meshes more than twice as wide as the clavae; clavae with a globular surface. Fig. 1/5–6 (p. 3).

UGANDA. Toro District: Ruwenzori, Bujuki Valley, March 1948, *Hedberg* 398! & Mobuku Valley, July 1952, *Ross* 537! & Ruwenzori, June–July 1968, *Hamilton* 739!
KENYA. Nanyuki District: Mt Kenya, Jan. 1971, *Hedberg* 4853!
DISTR. **U** 2; **K** 4; Congo (Kinshasa)
HAB. Along paths and small streams, on mud in bogs; 2900–4000 m

SYN. [*C. stagnalis* sensu Hedberg, A.V.P.: 127 (1957) p.p.; A. Robyns, F.C.B. 7: 351 (1958) p.p.; U.K.W.F. ed. 2: 82 (1994), *non* Scop.]

3. **C. keniensis** *Schotsman* in Bull. Mus. Hist. Nat. Paris, sér. 4, 10: 10 (1988). Type: Kenya, Turkana District, Murua Nysigar Peak, *Paolo* 977 (BR!, holo., EA, K!, PRE, iso.)

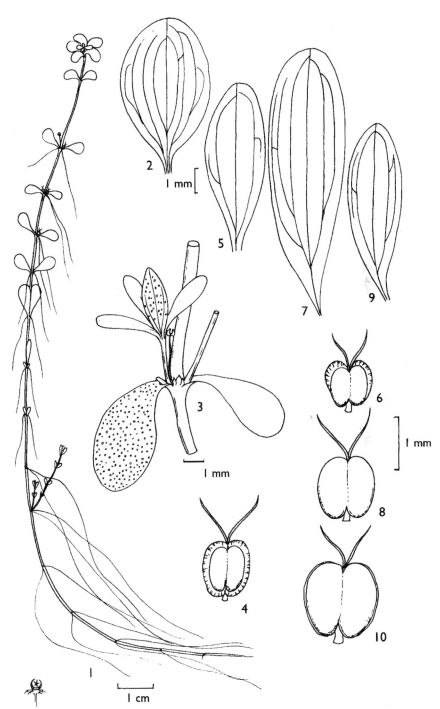

FIG. 1. *CALLITRICHE OREOPHILA* — **1**, habit; **2**, leaf; **3**, node with leaves and stamen; **4**, fruit. *C. ANISOPTERA* — **5**, leaf; **6**, fruit. *C. KENIENSIS* — **7**, leaf; **8**, fruit. *C. VULCANICOLA* — **9**, leaf; **10**, fruit. 1–2 from *Hedberg* 802, 3 from *Polhill* 437, 4 from *Bally* 9879; 5–6 from *Mabberley* 452 and after Schotsman; 7 from *Townsend* 2281, 8 from *Wesche* 116 and after Schotsman; 9–10 from *Hedberg* 604 and after Schotsman. Drawn by Henk Beentje.

Robust aquatic herb, with stems up to 40 cm long. Larger leaves spathulate to elliptic, up to 15 mm long and 4 mm wide, 5–8-veined; petiole 1(–3)-veined. Flowers either solitary or one male and one female in the same axil; bracts 2, about 1.5 mm long; stamens 3–4 mm long, after dehiscence up to 5 mm long; anthers 0.8 mm long. Fruit ± orbicular, ± 1.5 mm, unwinged. Pollen reticulum meshes not as wide as the clavae; clavae irregularly triangular with sharp angles. Fig. 1/7–8 (p. 3).

UGANDA. Mbale District: Elgon, Sasa trail in crater, Oct. 1996, *Wesche* 116!
KENYA. Turkana District: Murua Nysigar Peak, Sept. 1963, *Paulo* 977!; Mt Elgon, in the crater, May 1948, *Hedberg* 939!; Mt Kenya, Teleki Valley, July 1948, *Hedberg* 1763!
DISTR. U 3; K 2–4; not found elsewhere
HAB. In small ponds and streams; 2150–4250 m

SYN. [*C. stagnalis* sensu Hedberg, A.V.P.: 126 (1957) p.p.; U.K.W.F. ed. 2: 82 (1994), *non* Scop.]

4. **C. vulcanicola** *Schotsman* in B.J.B.B. 55: 294 (1985). Type: Kenya, Nanyuki District, Mt Kenya, W slope along the Burguret track, *Hedberg* 4403 (UPS!, holo., BR, EA, K!, iso.)

Prostrate and often mat-forming, sometimes submerged herb, with stems 5–15 cm long. Leaves spathulate or elliptic, up to 8(–10) mm long and 2–5 mm wide, usually 3-veined; petiole 1.5–3 mm long, 1-veined. Flowers solitary, bracteoles 1–2, ± 2 mm long, narrow; stamens 4–8 mm long, anthers ± 0.5 mm long; styles up to 1.5 mm long. Fruit 1.2–1.5 mm long and 1.6–2 mm wide, unwinged, but with distinctly edged margins. Pollen reticulum meshes as wide as the clavae; clavae obtusely triangular, surface almost smooth. Chromosome number: 2n=18 (counted on *Hedberg* 435, from Aberdare Mts and *Hedberg* 4403 from Mt Kenya (Hedberg & Hedberg 1988: 16)). Fig. 1/9–10 (p. 3).

UGANDA. Kigezi District: Muhavura, Oct. 1948, *Hedberg* 2117! & Mgahinga, Nov. 1954, *Stauffer* 789!; Toro District: Ruwenzori, Mijusi Valley, Mar. 1948, *Hedberg* 604!
KENYA. Aberdare Range, Kinangop, July 1948, *Hedberg* 1650!; Kiambu District: near junction of S Kinangop and Thiba roads, Apr. 1978, *Gilbert & Thulin* 1047!; Nanyuki District: Mt Kenya, Nov. 1967, *Hedberg* 4403!
DISTR. U 2; K 3, 4; Congo (Kinshasa)
HAB. Moist depressions in bog and grassland, along streams; 3000–4050 m

SYN. [*C. stagnalis* sensu Hedberg, A.V.P.: 126 (1957) p.p.; A. Robyns, F.C.B. 7: 351 (1958) p.p.; U.K.W.F. ed. 2: 82 (1994), *non* Scop.]

Callitriche deflexa *Hegelm.*, collected from Tanzania: Kilimanjaro, Layamungu Coffee Research Station, 4200 ft., *M. Bigger* 1236, is most likely an introduction. The species is native to South America and has been recorded from Portugal, Morocco and southern Africa, usually associated with cultivated areas (see Schotsman, op. cit.). It differs from the African species in being very small, with small leaves (2–4 mm long) and small fruits (0.5–0.7 mm).

INDEX TO CALLITRICHACEAE

Callitriche *L.*, 1
Callitriche anisoptera *Schotsman*, 2
Callitriche deflexa *Hegelm.*, 4
Callitriche keniensis *Schotsman*, 2
Callitriche oreophila *Schotsman*, 2
Callitriche stagnalis Scop., 2, 4
Callitriche stagnalis sensu auctt., 2, 4
Callitriche vulcanicola *Schotsman*, 4

No new names validated in this part

For Product Safety Concerns and Information please contact our EU representative GPSR@taylorandfrancis.com Taylor & Francis Verlag GmbH, Kaufingerstraße 24, 80331 München, Germany

Printed and bound by CPI Group (UK) Ltd, Croydon, CR0 4YY
01/05/2025
01858324-0001